Bibliographic information published by the German National Library:

The German National Library lists this publication in the National Bibliography; detailed bibliographic data are available on the Internet at http://dnb.dnb.de .

Imprint:

Print and binding: Books on Demand GmbH, Norderstedt Germany
ISBN: 9783668505285

This book at GRIN:

http://www.grin.com/en/e-book/370094/nutritional-benefits-of-goats-milk-major-components-and-medicinal-value

Ismail Tijjani Kabwanga, Cansu Altin

Nutritional Benefits of Goats' Milk. Major Components and Medicinal Value

GRIN Publishing

Nutritional Benefits of Goats’ Milk

Kabwanga Ismail Tijjani and Cansu Altin

Dairy Technology Dept., Faculty of Agriculture,
Ankara University

Abstract

Consumption of goats' milk is increasing worldwide due to the nutritional value and diversity of products produced from it. Goat milk is richer in Ca, Mg and P than cow and human milk. It's also rich in Medium Chain Triglycerides -MCT. Goat's milk a source of fat, protein, lactose, vitamins B6 (25%) more, 47% more vitamin A and 13 % more calcium than cow's milk. Goat's milk is vital in the management of allergies, gastro-intestinal disorders, CVD, osteoporosis, hypertension, colic, and antimicrobial protection. The objectionable odor produced by goats and the negative tradition beliefs demeans the advantages of goat milk consumption and the development of the sector, more so in developing countries. This review paper is focused at creating awareness of the nutritional values of goat's milk through highlighting the major components in the milk and its medicinal value using recent information and to recommend further investigation concerning nutritional and medicinal value of goat milk

Table of contents

1. Introduction

Goats are among the first animals to be domesticated by man over 10,000 years (Ulusoy ; Thornton 2010) The global increase in human population has lead to a high demand of milk more in the tropical developing countries (Da Silva et al. 2016; Cavicchioli et al. 2015).The worldwide increase in the consumption goats milk and its related products due to its organoleptic and great nutritional properties (Da Silva et al. 2016; Savoini et al. 2010; Ribeiro and Ribeiro 2010).The increased milk demand can be overcomed by increasing ruminant such as goats (Zhou et al. 2016). Goats also called a "Poor man's cow" and are an important source of milk (Iqbal et al. 2008) Goats can easily adapt to harsh climates making them an important component in the livestock industry more so to the poor small unit and marginal farmers (Zenebe et al. 2014)

There are about 861.9 million goats in the world (Zenebe et al. 2014). Over which 65% of the global goat populations are grazed in the developing countries of Asia and 29% in Africa (Bhattarai 2014; Zenebe et al. 2014). Goat milk was estimated to contribute about 2.3% of the worldwide milk production (Claeys et al. 2014).These numbers are likely to be much greater because there is much unreported amount of milk consumed from home consumption (Haenlein 2004).

There are varieties of products from goat's milk and other dairy animals. These products are nutritionally important as diet supplement and provide components necessary for human growth and health (Serhan, Mattar, and Debs 2016). Goat milk and its products are suggested to provide 3 major importance in human nutrition: (1) feeding more food insecure and malnourished individuals mostly in the developing countries than cow milk; (2) serving people affected with cow-milk allergies and gastro- intestinal disorders; and (3) meeting the gastronomic needs of connoisseur consumers whose market population is more in developed countries (Xi et al. 2015; Park 2012; Haenlein 2007) Thus, goat milk serves mainly three types of markets around the world: (a) home consumption, (b) specialty gourmet interests, and (c) medical needs (Park 2012)

However, goats are regarded as the most maligned domesticated animal in many parts of the world because of its offensive odor (Baranwal 2013; Griffiths 2010). However, good manufacturing practices (GMP) from feeding, milk and production processes helps to produce odor free goat milk hard to distinguish from cow milk both in odor and taste. Therefore, production of quality goat milk can help to reduce the prejudice against goat milk by consumers

(Haenlein 2007), good and no objectionable flavor, be free of spoilage bacteria, and contain legal minimum limits of all nutrients (Griffiths 2010).Therefore the objectives of this review paper to create awareness of the nutritional and medicinal values of goat milk and to recommend further investigation concerning nutritional and medicinal value of goat milk

2. Composition of goat milk and health benefits

Compositions of goat milk vary with diet, breed, individuals, parity, season, feeding, management, environmental conditions, locality, stage of lactation, and health status of the animal (Park 2012). Goat milk contains 3.8% fat, 3.4% protein, 4.1% lactose, 0.8% ash, 8.9 % SNF and 87% water. Goat milk differs from cow or human milk in having better digestibility, alkalinity, buffering capacity and certain therapeutic values in medicine and human nutrition (Bhattarai 2014; Park 2012).

Many studies show that some components in goat milk has more advantages of specific benefits in human nutrition and food security than other dairy species (Zhou et al. 2016; Cavicchioli et al. 2015).Goat milk is a rich source of protein, carbohydrate, lipid, vitamin and mineral. The superior digestibility of goat milk, the proper composition of fatty acids and its content of bioactive compounds (Zenebe et al. 2014).

2.1. Nutrition values and health benefits of goats' milk

2.1.1. Protein

The protein content in goat milk varies according to the breed, genetics, stage of milking, season, and feed (Park 2012; Park).There two major protein are; Casein (80%) s1, αs2, $\beta$and κ -caseins and the rest whey proteins -lactoglubulin and α –lactalbumin and the antigenicity of β -lactoglobulin can be partially eliminated by certain treatments (Silanikove et al. 2010; Serhan, Mattar, and Debs 2016; Zenebe et al. 2014). Goat milk has a small Casein micelles because of its higher concentrations of calcium and phosphorus (Zenebe et al. 2014).Goat milk luck genotype for CSN1S1expression hence making it low in α_{S1}-casein (4 g/L) compared to cow milk (7 g/L) because goats (Wei 2016; Tetens 2014; Prosser et al. 2008).

Goat milk has more β-casein solubilization and lower heat stability than cow's milk. The curd in goat-milk is weak hence easy digestibility in the gastrointestinal tract (Bhattarai 2014;

Horáčková et al. 2014). Goat milk is high in taurine amino acid than in cow milk (Ulusoy 2015 ; Rutherfurd et al. 2008; Belewu and Adewole 2009). Taurine is important for body growth and brain development, formation of bile salts, modulation of calcium flux and the stabilization of membranes as an osmoregulation elimination of toxic substances. Deficient of Taurine in humans results in retarded growth, epilepsy and cardiomyopathy (Ulusoy 2015 ; Zenebe et al. 2014; Silanikove 2008).

The bioactive peptides such as immunoglobulin, lactoferrin, lactoperoxidase, folate binding protein and more recently, α-lactalbumin and β-lactoglobulin are Carrier of retinol, fatty acids and triglycerides and provide body passive immunity; Anti-carcinogenic activity, treatment of chronic stress-induced diseases; Anticarcinogenic; Antioxidant activity; Antimicrobial activity; Antifungal activity; Anti-thrombotic activity and good for growth and development (Wei 2016; Rachman, Maheswari, and Bachroem 2015; Hernández-Ledesma, Ramos, and Gómez-Ruiz 2011) and immuo-stimulating peptide can be generated from caprine –caseins (Zenebe et al. 2014). Goat milk has Antihypertensive properties as its a good source of ACE-inhibitory peptides resulting to goat milk caseins hydrolysis hence regulation blood pressure (Kabwanga Ismail Tijjani 2016). Goat milk provides complete protein with all essential amino acids with low fat content and mucus forming components like cow milk (Díaz-Castro et al. 2010).

2.1.2. Milk fats

Goat milk is a good source of fats, which is nutritionally important and determinant for price milk (Haenlein 2007; Zenebe et al. 2014). Tricacylglycerol over 98% (Kompan and Komprej 2012) and simple lipids such as monoacylglycerol and complex lipids such as phospholipids, sterols hydrocarbs and cholesterol esters are among the largest fats in goat milk. Goats-milk has great portion of conjugated linoleic acid(CLA) which helps to boost body immunity by stimulating immunoglobulns, cytokines and prostaglandins immunity mediators (Zenebe et al. 2014), anticarcinogenic, antiatherogenic, antidiabetic, immunomodulation and antiadipogenic (Kompan and Komprej 2012; Savoini et al. 2010).

Studies show that polyunsautrated fat acids (PUFA) such as eicosapentaenoic acid (EPA, C20:5) and docosahexaenoic acid (DHA, C22:6) provide anti- inflammatory properties and early infant neuronal development (Savoini et al. 2010).

Medium chain triacylglycerol (MCTs) that are directly nutrient absorption in the ileum without degradation, good for body energy production and antimicrobial protection (Ceballos et al.

2009; Zenebe et al. 2014; López-Aliaga et al. 2010). MCTs short and medium chain fatty acids (FFA) like butyric lauric, caprylic, linoleic ,palmitic are much found in great quantity in goat-milk with some (FFA) like caproic and capric acid deliver their names from goat (Bergillos-Meca et al. 2015; Ceballos et al. 2009; Haenlein 2007). The lipid globule membranes in goat milk are smaller and more evenly distributed for easier digestion in human (Bergillos-Meca et al. 2015; Zhou et al. 2016; Baranwal 2013; Morgan 2012), the low fat content of goat-milk and its ability to neutralize acids and toxins makes it a therapeutic product for humans (Pacinovski et al. 2015).

2.1.3. Milk carbohydrate

Goat milk has more lactose content than cow milk. Lactose helps in intestinal absorption of calcium, magnesium and phosphorous and also the utilization of vitamin D (Zenebe et al. 2014; Ceballos et al. 2009). However, oligosaccharides, glycoproteins, glycopeptides, and nucleotides are found in small quantity. Goat milk is high in lactose-derived oligosaccharides compared to cow milk (Horáčková et al. 2014). Oligosaccharides act as prebiotic and anti-inflammatory effects in induced colitis and inflammatory bowel disease control. This is achieved through their increased production of butyrate and the reduction of pro-inflammatory bacterial species by inhibiting their adhesion to the epithelial membrane, reducing bacterial translocation and hence promoting selective growth of beneficial Lactobacillus and Bifidobacteria species (Ulusoy 2015; Viverge, Grimmonprez, and Solere 1997).

Studies on rats with colitis in Spain showed that, goat milk oligosaccharides has anti-inflammatory effects, hence useful in the management of inflammatory bowel disease (IBD) and promotes recovery of damaged colonic mucosa (Zenebe et al. 2014; Daddaoua et al. 2006). Goat-milk is considered as food due to its easier fat digestibility as well as the lower concentration of functional allergenic proteins and reduced allergenicity of goat milk in relation to cow milk (Savoini et al. 2010; Gomes et al. 2015).

Milk Vitamin:

Goats convert all β-carotenes from foods into vitamin A. This makes vitamin A to be more in goat's milk than in cow milk (Conesa et al. 2008). Vitamin B6 and vitamin D content is low in cow milk as well as in goat milk (Horáčková et al. 2014). The vitamin A in goat milk similar to that of human milk. Vitamin A boosts body immune responses and antibody responses. Vitamin D in milk is recommended in the dietary management of osteoporosis, diabetes and cancer. Goat milk is high in Vitamin C, which is an antioxidant and has antiviral properties than in cow milk (Geissler 2011).

2.1.4. Milk mineral

Goat milk has more potassium, calcium, chloride, phosphorus, selenium, zinc and copper than cow milk (Díaz-Castro et al. 2010). Potassium is important for the acid-base balance, muscles functioning, nerves and kidney health. Chloride is good for liver function, maintains fluid balance, blood pH and osmotic pressure. Calcium is of importance for strong bone structure and blood coagulation. Selenium is good for cell protection against free radicals. Zinc is high in goats milk and its used for healthy skin, wound healing and it is also has antioxidative properties and a cofactor for the antioxidant enzyme, protein production and it also useful in hormone insulin regulation for carbohydrates breakdown (Griffiths 2010; Díaz-Castro et al. 2010).

2.2. Human health qualities of goat milk

The nutritional and therapeutic values of goat milk makes special for human consumption (Ribeiro and Ribeiro 2010).

2.2.1. Cardiovascular Diseases

Cardiovascular disease (CVD) such as; coronary heart disease, high blood pressure, arrhythmias and atherosclerosis are the world leading cause of death (Kabwanga Ismail Tijjani 2016). The etiology of CVD's are; Smoking, diet, physical inactivity, high blood pressure, dyslipidemia, diabetes and obesity among others (Zenebe et al. 2014).

The medium-chain triglycerides (MCT) such as the caproic, caprylic and capric fatty acids high in goat-milk showed a lowering effect on plasma cholesterol in rat limited cholesterol deposition in rat tissues (Kompan and Komprej 2012). Also MCT act as anti-atherogenic and antioxidative (Kompan and Komprej 2012). Consuming goat milk stimulates the release of nitric oxide (NO) thus promoting vasodilatation and exerts a cardio-protective and anti-atherogenic affect (Savoini et al. 2010; Claeys et al. 2014). The anti-oxidation are important in management of CVD risk factors like body overweight, type 2 diabetes mellitus (T2DM); and prevents hyperhomocysteinemia (Kabwanga et al. 2016). Goat milk is rich in carnitine content that increases the rate of ß-oxidation in mitochondria hence inhibiting manifestation of CVDs

(Zenebe et al. 2014).

2.2.2. Treatment of gastrointestinal diseases

Inflammatory bowel disease (IBD) includes conditions like, Crohn´s disease and ulcerative colitis. The ulcerative colitis affects the large intestine at the mucosal level, whereas Crohn´s disease is characterized by transmural inflammation in the gastrointestinal tract (Daddaoua et al. 2006). The oligosaccharides in goat milk have an anti-inflammatory effect, body weight reduction, increase the colon size and extension of necrotic lesions. They also important in diarrhea and bloody stools management (Ulusoy 2015 ; Zenebe et al. 2014)

2.2.3. Allergy

Milk is a common source of food allergies, with high prevalence in children but though never last for long (Russell et al. 2011). Studies show that, goat milk has less allergenic risk due to its different in protein structure (casein micelle) components hence an alternative to cow milk (Ribeiro and Ribeiro 2010; Tetens 2014). The sour curd formed in the stomach as a result of the much lower content of a peculiar type of casein, αs-1 casein makes goats milk simpler to be digested including the lactose portion hence less transit in the colon hence minimizing lactose intolerance (Zenebe et al. 2014; do Egypto et al. 2013).

Intake of goat milk has over 30 - 40% reduction of allergic cases than cow milk (Russell et al. 2011; Ballabio et al. 2011) . There is few data regarding goat-milk protein and allergenicity. This makes it complicated to predict the incidence and severity of adverse reaction to goat milk protein. Also is insufficient evidence suggesting a lower incidence of allergic reactions as a result of goat milk-based infant formula compared with cow milk-based infant formula foods. However, it's important to note that, use of goat milk protein as an alternative for cow milk protein in infant formula intended for cow milk allergic risk infants cannot be considered safer unless more clinical trials are carried to detail more evidence of goat milk protein as less allergy free than cow milk (Wei 2016; Zhou et al. 2016).

2.2.4. Bone health

Generally milk is a rich source of calcium, which promotes bone health. Consumption of 200ml of goat's milk gives about 29% of the UK dietary reference value of calcium for an adult. Deficiency in calcium leads to osteoporosis (Baranwal 2013). Study showed a reduced risk of fracture (34.8%) in children who consumed milk compared to (13%) in children who didn't take milk (Baranwal 2013).

2.2.5. Antimicrobial Properties

Milk proteins are rich in antimicrobial peptides like lactoferrin. Lactoferrins bind-iron whose main role is to iron transport. This provides antimicrobial, immune response and exhibition of antioxdation properties (Kompan and Komprej 2012).

There is rapid decrease of pH values of goat milk during fermentation. Some authors indicated that the higher fermentation activity of lactic acid bacteria in goat milk is due to its specific composition and structure (Zenebe et al. 2014). Many studies show that fermented milk with probiotics inhibits gram -negative bacteria. The mechanism in which fermented goat milk prevents pathogenic microorganisms is not clearly known. Its suggest that spoilage and pathogenic microbe in *Lactobacillus* strains are a vital dairy culture starter variety for foods safety (Ulusoy 2015; Zenebe et al. 2014).

3. Conclusion

Goats' milk products are very important in human health because of their nutritive and therapeutic values. Goat milk has unique nutritional values superior to other milk source species such as cows. It's easy to digest, good composition of fatty acids, protein and its content of bioactive compounds makes it more of medicinal value than a mere food. Its beneficial effects ranges from malabsorption disorders and inflammatory bowel diseases control reduce the risk of cardiovascular disease due to its ant-oxidative properties among others. However, the offensive order produced by goats that is later transferred to the milk makes many people to reject its consumption. There is need to develop an effective good manufacturing processes such as feeding-milking time, disease prevention, proper storage conditions and heat treatments that help to get ride of that smell and also preserve its nutritional value. It is of paramount to sensitize the public to increase goat's milk and its derivatives for body health.

4. Bibliography

Ballabio, C, S Chessa, D Rignanese, C Gigliotti, G Pagnacco, L Terracciano, A Fiocchi, P Restani, and AM Caroli. 2011. 'Goat milk allergenicity as a function of α S1-casein genetic polymorphism', *Journal of dairy science*, 94: 998-1004.

Baranwal, Deepika. 2013. 'The Benefits of Consuming Goat's Milk', *Trends in Biosciences*, 6: 513-15.

Belewu, MA, and AM Adewole. 2009. 'Goat milk: A feasible dietary based approach to improve the nutrition of orphan and vulnerable children', *Pakistan Journal of Nutrition*, 8: 1711-14.

Bergillos-Meca, Triana, Carmen Cabrera-Vique, Reyes Artacho, Miriam Moreno-Montoro, Miguel Navarro-Alarcón, Manuel Olalla, Rafael Giménez, Isabel Seiquer, and Maria Dolores Ruiz-López. 2015. 'Does Lactobacillus plantarum or ultrafiltration process improve Ca, Mg, Zn and P bioavailability from fermented goats' milk?', *Food chemistry*, 187: 314-21.

Bhattarai, Rewati Raman. 2014. 'Importance of goat milk', *Journal of Food Science and Technology Nepal*, 7: 107-11.

Cavicchioli, VQ, TM Scatamburlo, AK Yamazi, FA Pieri, and LA Nero. 2015. 'Occurrence of Salmonella, Listeria monocytogenes, and enterotoxigenic Staphylococcus in goat milk from small and medium-sized farms located in Minas Gerais State, Brazil', *Journal of dairy science*, 98: 8386-90.

Ceballos, Laura Sanz, Eva Ramos Morales, Gloria de la Torre Adarve, Javier Díaz Castro, Luís Pérez Martínez, and María Remedios Sanz Sampelayo. 2009. 'Composition of goat and cow milk produced under similar conditions and analyzed by identical methodology', *Journal of Food Composition and Analysis*, 22: 322-29.

Claeys, WL, Claire Verraes, Sabine Cardoen, Jan De Block, André Huyghebaert, Katleen Raes, Koen Dewettinck, and Lieve Herman. 2014. 'Consumption of raw or heated milk from different species: An evaluation of the nutritional and potential health benefits', *Food Control*, 42: 188-201.

Conesa, Celia, Lourdes Sánchez, Carmen Rota, María-Dolores Pérez, Miguel Calvo, Sebastien Farnaud, and Robert W Evans. 2008. 'Isolation of lactoferrin from milk of different species: calorimetric and antimicrobial studies', *Comparative Biochemistry and Physiology Part B: Biochemistry and Molecular Biology*, 150: 131-39.

Da Silva, Fabiana Fernanda Pacheco, Vanessa Biscola, Jean Guy LeBlanc, and Bernadette Dora Gombossy de Melo Franco. 2016. 'Effect of indigenous lactic acid bacteria isolated from goat milk and cheeses on folate and riboflavin content of fermented goat milk', *LWT-Food Science and Technology*, 71: 155-61.

Daddaoua, Abdelali, Victor Puerta, Pilar Requena, Antonio Martínez-Férez, Emilia Guadix, Fermín Sánchez de Medina, Antonio Zarzuelo, María Dolores Suárez, Julio José Boza, and Olga Martínez-Augustin. 2006. 'Goat milk oligosaccharides are anti-inflammatory in rats with hapten-induced colitis', *The Journal of nutrition*, 136: 672-76.

Díaz-Castro, J, S Hijano, MJM Alférez, I López-Aliaga, T Nestares, M López-Frías, and MS Campos. 2010. 'Goat milk consumption protects DNA against damage induced by chronic iron overload in anaemic rats', *International dairy journal*, 20: 495-99.

do Egypto, Rita de Cássia Ramos, Bárbara Melo Santos, Ana Maria Pereira Gomes, Maria João Monteiro, Susana Maria Teixeira, Evandro Leite de Souza, Carlos José Dias Pereira, and Maria Manuela Estevez Pintado. 2013. 'Nutritional, textural and sensory properties of Coalho cheese made of goats', cows' milk and their mixture', *LWT-Food Science and Technology*, 50: 538-44.

Gomes, LC, CR Alcalde, GT Santos, AC Feihrmann, BSL Molina, PA Grande, and AA Valloto. 2015. 'Concentrate with calcium salts of fatty acids increases the concentration of polyunsaturated fatty acids in milk produced by dairy goats', *Small Ruminant Research*, 124: 81-88.

Griffiths, Mansel W. 2010. *Improving the Safety and Quality of Milk: Improving Quality in Milk Products* (Elsevier).

Haenlein, GFW. 2004. 'Goat milk in human nutrition', *Small Ruminant Research*, 51: 155-63.

Haenlein, G. 2007. 'About the evolution of goat and sheep milk production', *Small Ruminant Research*, 68: 3-6.

Hernández-Ledesma, Blanca, Mercedes Ramos, and José Ángel Gómez-Ruiz. 2011. 'Bioactive components of ovine and caprine cheese whey', *Small Ruminant Research*, 101: 196-204.

Horáčková, Šárka, Pavla Sedláčková, Marcela Sluková, and Milada Plocková. 2014. 'Influence of Whey, Whey Component and Malt on the Growth and Acids Production of Lactobacilli in Milk', *Czech J. Food Sci. Vol*, 32: 526-31.

Iqbal, A, BB Khan, M Tariq, and MA Mirza. 2008. 'Goat-A potential dairy animal: Present and future prospects', *Pakistan Journal of Agricultural Sciences (Pakistan)*.

Kabwanga Ismail Tijjani, Ceren Akal and Atila Yetisemiyen. 2016. 'Role of Milk, Dairy Products and Milk Components Used in the Management of Metabolic Syndrome', *Research Journal of Agriculture and Forestry Sciences*, Vol. 4(9), 14-20.

Kompan, Dragomir, and Andreja Komprej. 2012. 'The effect of fatty acids in goat milk on health', *Milk production-An up-to-date overview of animal nutrition, management and health. Chaiyabutr, N.(ed) InTech, Croatia*: 1-26.

López-Aliaga, Inmaculada, Javier Díaz-Castro, Ma José M Alférez, Mercedes Barrionuevo, and Margarita S Campos. 2010. 'A review of the nutritional and health aspects of goat milk in cases of intestinal resection', *Dairy science & technology*, 90: 611-22.

Morgan, D., Gunneberg, C., Gunnell, D., Healing, T.D., Lamerton, S., Brooks, J., 580 Martinez, B., Stratton, J., Bianchini, A., Krokstrom, R. and Hutkins, R. 2012. 'Medicinal properties of goat milk', *J. Dairy Goat.*

Pacinovski, Nikola, Gordana Dimitrovska, Ljupčo Kočoski, Goce Cilev, Mirjana Menkovska, Biljana Petrovska, and Apostol Pacinovski. 2015. 'Nutritive advantages of goat milk and possibilities of its production in Republic of Macedonia', *Maced. J. Anim. Sci*, 5: 81-88.

Park, Young W. 2012. 'Hypo-allergenic and Therapeutic Significance of Goat Milk'.

Park, YW. 2012. "Goat milk and human nutrition." In *First Asia Dairy Goat Conference*, 31.

Prosser, Colin G, Robert D McLaren, Deborah Frost, Michael Agnew, and Dianne J Lowry. 2008. 'Composition of the non-protein nitrogen fraction of goat whole milk powder and goat milk-based infant and follow-on formulae', *International journal of food sciences and nutrition*, 59: 123-33.

Rachman, Agus Bahar, Rarah RA Maheswari, and Mirnawati S Bachroem. 2015. 'Composition and Isolation of Lactoferrin from Colostrum and Milk of Various Goat Breeds', *Procedia Food Science*, 3: 200-10.

Ribeiro, AC, and SDA Ribeiro. 2010. 'Specialty products made from goat milk', *Small Ruminant Research*, 89: 225-33.

Russell, DA, RP Ross, GF Fitzgerald, and C Stanton. 2011. 'Metabolic activities and probiotic potential of bifidobacteria', *International journal of food microbiology*, 149: 88-105.

Rutherfurd, Shane M, Paul J Moughan, Dianne Lowry, and Colin G Prosser. 2008. 'Amino acid composition determined using multiple hydrolysis times for three goat milk formulations', *International journal of food sciences and nutrition*, 59: 679-90.

Savoini, G, A Agazzi, G Invernizzi, D Cattaneo, L Pinotti, and A Baldi. 2010. 'Polyunsaturated fatty acids and choline in dairy goats nutrition: Production and health benefits', *Small Ruminant Research*, 88: 135-44.

Serhan, Mireille, Jessy Mattar, and Liliane Debs. 2016. 'Concentrated yogurt (Labneh) made of a mixture of goats' and cows' milk: Physicochemical, microbiological and sensory analysis', *Small Ruminant Research*, 138: 46-52.

Silanikove, N, G Leitner, U Merin, and CG Prosser. 2010. 'Recent advances in exploiting goat's milk: quality, safety and production aspects', *Small Ruminant Research*, 89: 110-24.

Silanikove, Nissim. 2008. 'Milk lipoprotein membranes and their imperative enzymes.' in, *Bioactive Components of Milk* (Springer).

Tetens, Inge. 2014. "EFSA NDA Panel (EFSA Panel on Dietetic Products, Nutrition and Allergies), 2014. Scientific Opinion on the essential composition of infant and follow-on formulae." In.: Europen Food Safety Authority.

Thornton, Philip K. 2010. 'Livestock production: recent trends, future prospects', *Philosophical Transactions of the Royal Society B: Biological Sciences*, 365: 2853-67.

Ulusoy Beyza Hatice. 2015. 'Nutritional and Health Aspects of Goat Milk Consumption'.

Viverge, Danièle, Louis Grimmonprez, and Maryse Solere. 1997. 'Chemical characterization of sialyl oligosaccharides isolated from goat (Capra hircus) milk', *Biochimica et Biophysica Acta (BBA)-General Subjects*, 1336: 157-64.

Wei, Ruobin. 2016. 'The Effects of Consuming Goat's Milk Infant Formula on the Growth Outcomes of Infants and Children Compared to Cow's Milk Infant Formula'.

Xi, Meili, Yuqing Feng, Qiong Li, Qinnan Yang, Baigang Zhang, Guanghui Li, Chao Shi, and Xiaodong Xia. 2015. 'Prevalence, distribution, and diversity of Escherichia coli in plants manufacturing goat milk powder in Shaanxi, China', *Journal of dairy science*, 98: 2260-67.

Zenebe, Tilahun, Nejash Ahmed, Tadele Kabeta, and Girma Kebede. 2014. 'Review on Medicinal and Nutritional Values of Goat Milk', *Academic Journal of Nutrition*, 3: 30-39.

Zhou, Fengyan, Qing Yang, Chuzhao Lei, Hong Chen, and Xianyong Lan. 2016. 'Relationship between genetic variants of POU1F1, PROP1, IGFBP3 genes and milk performance in Guanzhong dairy goats', *Small Ruminant Research*.